Agile Decisions

Michael Nir

Agile Decisions

Driving Effective Agile Decisions in Business

A humoristic practical approach to understanding why decisions are so difficult and what can be done about it

Agile Decisions

Michael Nir

Agile Decisions

Agile Decisions

Driving Effective Agile Decisions

in Business

Michael Nir

Agile Decisions

Preface

When I was young I loved reading Mad magazine. I guess this more or less uncovers my age.

Recently, Al Feldstein, who was eulogized as the soul behind Mad magazine passed away at the age of 88. Mad magazine has impacted the way I write and the way I present information in lectures, workshops and while coaching: *always take it with a grain of salt, and make sure you're not overly serious.*

For me it was always about the: *Lighter side of...*

I am dedicating this book as a homage to Al.

I think he would've liked my thumbs-up examples that I'm using in this book – for example:

*If for example you communicate ahead of time, to your family members, that you **never carry gifts** to other members and relatives, and you're able to follow through on this rule, you are actually better off packing your luggage. You might also be ostracized and banned from the next thanksgiving; such is the price you pay for packing light.*

Agile Decisions

For your enhanced reader experience, we use the following interactive tools in the book:

> **Remember** – remember scrolls summarize concepts discussed to assist in information retention

Thumbs up – these are there for comic relief or side-ideas when needed....

Reflect – reflection clouds invite you to invest deeper thinking of specific ideas and insights

I truly hope you like this book!

Agile Decisions

About the Author

Michael Nir - President of Sapir Consulting US LLC - PMP, Scaled Agile Consultant - has been helping clients overcome business challenges and achieve their potential for over 16 years. He is passionate about Gestalt theory and practice, which complements his civil and industrial engineering background (M.Sc. and B.Sc.) and contributes to his understanding of individual and team dynamics in business. Michael authored bestsellers on Influencing, Agile, Teams, and Leadership. His experience includes significant know-how in the telecoms, hi-tech, banking, R&D environments and petrochemical & infrastructure industries. He develops creative and innovative solutions in Agile project and product management, process improvement, leadership, and team building.

Michael's professional background is analytical and technical; however, he has a keen interest in human interactions and behaviors. He holds two engineering degrees from the prestigious Technion Institute of Technology: a Bachelor of civil engineering and Masters of Industrial engineering. He has balanced his technical side with the extensive study and practice of Gestalt Therapy and "Instrumental Enrichment," a philosophy of mediated learning. In his consulting and training engagements, Michael combines both the analytical and technical world with his focus on people, delivering unique and meaningful solutions, and addressing whole systems.

What readers say about this book:

> *".....*It is what makes this decision making book **so approachable and powerful**. i can endorse the examples and solutions presented. ...."

> *".....This weekend I received a copy of Michael Nir's ninth book Agile Decisions, practical decision making business guide. A small book which I read during my flight from Amsterdam to Athens. It's an easy to read book about an interesting and difficult topic. It's all about decision making. Michael uses from time to time simple icons to show you a reminder, to make a reflection or give a thumbs-up with an easy to understand example.* **will help you to reflect on your own decision experiences and gives you some guidelines to support your future decision***...."*

> *".....*I warmly **recommend** this book. It's useful and entertaining as well....."

"…..*Agile Decisions* is **truly a humoristic practical approach** to understanding why decisions are so difficult and what can be done about it.

The author write that: I really hope I've succeeded in making this decision making book an enjoyable and interesting adventure for you. The many **examples make it worth the time to investigate more into the practical and fundamentals**.

The five guidelines make sense, I will definitely think about it on my next summer holiday…**all in all a fun to read** and guiding book…."

Agile Decisions

Contents

Introduction 2

Complexity is underrated 6

Embrace Complexity 8

Can you think of a complex problem for a long period of time?
....11

Imagine that you're a travelling salesperson.............13

The good old school days......................................25

Packing to go the Caribbean...................................28

Summary..33

The sushi was poisonous 36

The right people for the job 40

Golden arches use computers..................................48

Japanese thinking 59

Taking off: Pilots with Food Poisoning.......................65

Building Pyramids..74

They who have greater position power 77

They who yell louder...83

Cognitive biases 90

Flood management on the Nile 95

Fusion, Packing Nothing to Moscow 104

When everything is important, nothing is *105*

Wearing the same shirt five days in a row................................*112*

Five Practical Guidelines for Agile Decisions............................*116*

    Simple Local Rules 117

    Strategic top-down rules 120

    Visual Problem Presentation..127

    Realignments feedback mechanism.....................................131

    Enforce consistency through publicity 137

Decisions, the road not taken **149**

Agile Decisions, satisfying rather than optimizing..................*150*

Focus and Limiting Options...*152*

Focus: Need, top-down rule *156*

What is unique about this guide..*160*

What's next and final words **163**

Michael Nir

Introduction

This book is about decision making.

Decision making is serious business; hence, a book about decisions has to be, and usually is, pompous.

While I discuss complex decision problems in this book, I aim to make it light and fun with many examples and anecdotes.

We think that we fail in decisions because of our own faults. Many recent popular science books have gleefully shown how irrational we are in making decisions. However, that is not the main reason we fail in our decisions. The true challenge stems from our failure to grasp complexity in most everyday life problems.

In this book, we discover what happens when an airline pilot has food poisoning from sushi in Bangkok, why flying is such a hassle, and whether we will be kidnapped by the NSA. We also visit ancient Egypt and learn what the meek engineer does to prevent a flood in the capital and how it relates to selecting which items you're packing on your next trip to the Caribbean.

Practically, we offer process **guidelines** to successfully solve complex problem decisions in an Agile approach.

I truly hope that you find this book interesting and invigorating.

Sincerely,

Michael, 2014

Agile Decisions

CHAPTER
ONE

Complexity is underrated

Complexity is underrated

Why should we care about complexity, and how does it lead to bad decisions?

We really misunderstand and underestimate complexity in and around us. Many times, we mistake complexity for complication. We think that by having an orderly procedure we are able to solve a complicated problem, make a decision, and bask in the light of optimal results.

I was listening to a keynote that a colleague from a well-known business school was delivering. He discussed the flight patterns of aircrafts over Europe on a given day. There are thousands of flights every day. Routing them over the continent is extremely difficult; it is a complicated task of coordination and communication.

There is a well-thought protocol for deciding on flight routing. This is an example of a **complicated** endeavor. However, it is not an example of complexity in decision making.

In this book, I describe complex problems and how they affect decisions, yielding less than optimal results. These problems, on top of being difficult, are actually unsolvable in a reasonable amount of time. Many times, we take shortcuts to handle them. Often these shortcuts provide adequate results ... other times they don't.

Embrace Complexity

Our problems in business and life are complex and pose a challenge to our cognitive abilities. That said, most of the decisions answering these problems are based on rules of thumb that we've developed through experience to solve complex problems.

In order to understand the problems and our decision-making shortcuts and rules of thumb in business and outside, we first have to understand more about complexity.

Take a minute now to think about the word "complexity" and what it means to you. Write down a definition for complexity on paper.

You might recall that you once took a course about complexity at a university, especially if you majored in computer science or other engineering laden degrees.

If you did, the professor probably said that complex problems are such that we are unsure what the solution looks like.

 I noticed that most people remember very little from what they've studied at school, college, or university. They do remember the parties though, both real and imaginary.

I was searching for a definition of complexity to add to this book, but noted physicist professor **Neil Fraser Johnson** admits that "even among scientists, there is no unique definition of complexity—and the scientific notion has traditionally been conveyed using particular examples..."

I will shortly be introducing complex problem examples. Yet, I ask that you take a minute and write down your personal definition for complexity and a complex problem.

Reflect: What is your definition for complexity? Can you think of a complex problem you're facing?

I noticed that people use the word **"complicated"** to describe business problems and **"complex"** to describe people's personalities. Maybe this proves that we correctly define the concept of complexity because personalities are difficult to understand.

Can you think of a complex problem for a long period of time?

The truth is that many times when we are facing complex decision-making problems in business, and in life in general, we tend to substitute the problem with an easier one.

Cognitively, it is very difficult to linger on problems that are complex. It taxes our cognitive mental systems. Our cognitive abilities and powers are limited and require resources to function. We don't like to think too hard for too long.

Following Nobel laureate Daniel Kahneman's research:

> *System 2, in Kahneman's scheme, is our slow, deliberate, analytical, and consciously effortful mode of reasoning about the world. System 1, by contrast, is our fast, automatic, intuitive, and largely unconscious mode. It is System 1 that detects hostility in a voice and effortlessly completes the phrase "bread and..." It is System 2 that swings into action when we have to fill out a tax form or park a car in a narrow space.*
> *More generally, System 1 uses association and metaphor to produce a quick and dirty draft of*

reality, which System 2 draws on to arrive at explicit beliefs and reasoned choices. System 1 proposes. System 2 disposes. So System 2 would seem to be the boss, right? In principle, yes. But System 2, in addition to being more deliberate and rational, is also lazy. And it tires easily. Too often, instead of slowing things down and analyzing them, System 2 is content to accept the easy but unreliable story about the world that System 1 feeds to it. "Although System 2 believes itself to be where the action is," Kahneman writes, "the automatic System 1 is the hero of this book." System 2 is especially quiescent, it seems when your mood is a happy one.

Thinking Fast and Slow, D. Kahneman, The New York Times Book Review

We can trace our preference to limit the usage of cognitive power to conserve energy all the way back to our primate ancestors.

From a survival standpoint, it is better to make rapid decisions that enable us to flee from danger, find our next meal, and locate a suitable mate.

For this reason, I want you to linger on a complex problem. Remembering your definition for a complex problem, concentrate on a complex problem that you are, at present, facing.

It could be a business decision-making problem that is complex and involves many variables, constraints, and assumptions. You might be unsure concerning the different parameters of the problem and the many possible solutions for it.

Invest the necessary cognitive mental effort to fully grasp this problem.

Reflect: Linger for a few minutes contemplating the complex decision problem that you've thought of.

Imagine that you're a travelling salesperson.

Are you done thinking of your complex decision problem?

Remember it and compare with the complex problem I will be introducing in the following pages, giving you a taste of a decision problem that seems very simple, yet the optimal solution is extremely difficult to locate.

It is known as the "**traveling salesperson** problem."

The traveling salesperson problem is a nondeterministically polynomial (NP). We will discuss what NP complexity implies later. It will be easier to understand through an example.

Imagine that you are a traveling salesperson. You reside peacefully in a nice suburban house in the outskirts of Seattle. You're tasked with selling electronic appliances to various businesses in cities across Washington.

You can visit several possible locations. Your itinerary includes the following limit: You can visit every location only once. This coincides with reasonable travel restrictions for limiting and minimizing business travel costs.

Michael Nir

To reiterate: The one and only constraint is that every travel destination or city can be visited once in the entire route plan.

Naturally, every location has a certain potential value, i.e., revenues or profit.

The decision you need to make: Taking into consideration the travel costs from and to every location, what is the most lucrative route you can plan?

Other interesting decisions might be:

- ✓ What is the least travel time duration, e.g., are you able to complete the route and be home for Thanksgiving?

- ✓ What is the best solution in terms of the least distance traveled altogether?

- ✓ What is the most lucrative route to travel in terms of client availability?

These are just a few questions related to this problem.

Reflect on this decision problem for a few minutes.

How many possible solutions can you think of?

Actually, it seems that for 10 cities, this problem isn't very complex. Therefore, you decide to pack your carry-on, say goodbye to your spouse and family, and head to Seattle-Tacoma International Airport for the next flight to—?

Your company has bought you an open ticket for the trip. You are pondering your options as you reach the check in, looking for the first destination. You can travel to any of the 10 cities.

As you're taking off your shoes and your pouch, handing over your travel card, unpacking your iPad and laptop, and placing them in the bin for the security check, you think to yourself that maybe it is best to just pick the first travel location according to—

Heym just a minute, that was you buzzing as you passed through the security gate—

Officer, please be careful with those groping hands, those aren't my—

Still, the problem lingers in your head as you're picking up your belongings, hurried by the next couple in line who are traveling on their second honeymoon to Hawaii ... but I digress.

You decide that you will travel to the **least expensive** destination. You also **decide** that this will be your **general rule for making decisions**. You think to yourself that since the flight to Spokane, Washington, costs only $100, which is the least expensive option, it should be the next travel destination.

Once in Spokane you **decide** to look for the cheapest flight to one of your remaining nine destinations.

After following your rule of thumb, you complete the journey and calculate that the total travel costs have been $7,990.

You're proud of yourself.

If it weren't for that flight leg from **Yakima** to **Kennewick**, for which you paid an exuberant $2,560, you would have totaled around $5,500. Still, you are quite happy with your route, thinking that you managed to save on travel costs. You are also happy with your decision-making algorithm for figuring out the flight plan.

You found a decent solution for solving the travelling salesperson problem. Or have you?

Remember that the goal was to find the least expensive route. Does your rule of thumb for selecting the next cheapest travel destination actually provide the cheapest route **overall**?

Wait a minute. Am I hinting that it isn't the cheapest route?

It just makes logical sense that it is, doesn't it?

Here is the thinking pattern you developed: There is a complex decision problem and we are not certain how many options there are to solve it. There is limited time to decide, so we figure that by selecting each time the next least expensive route we also pay the least overall.

Actually, this thinking approach is what makes business decision rules of thumb so alluring and confusing. Often, when businesses face a challenging problem, they think that by breaking it down to many small elements, or rather many, many small problems, they can solve the big one by solving all the small ones.

This reminds me of the mechanistic view and project  management. In projects, there is a conviction that if we break down the big task to many small ones and estimate and schedule all the small tasks and aggregate, the result is a robust, comprehensive plan.

As much as it doesn't work in project management, the piecemeal approach doesn't necessarily provide an optimal solution for the traveling salesperson decision problem.

Here's the reason:

Let's assume that Spokane is the eighth city.

The travel from **Spokane to Yakima** is $500, the travel from **Spokane to Kennewick** is $600.

Naturally, we select to fly to **Yakima**.

Once in Yakima, we have no choice but to fly to our last remaining city, which is Kennewick, for which we unfortunately pay an exuberant $2,560.

However, had we first flown to Kennewick and paid only $100 more, we would have found the weekend deal flight and paid $750 for the flight from Kennewick to Yakima.

So, in total, by deciding on our next step without considering future travels, we saved $100 on the immediate flight, but paid $1,810 on the next flight.

Piecemeal solutions do not necessarily provide the best solution, though they are easy on the mind. Businesses fail to acknowledge it.

How can we reach the optimal plan—the best travel route in terms of price?

It is a good idea to calculate the overall number of routes that we have.

Traveling to 10 locations/cities in Washington, the number of route options is easily calculated by 10!

Okay, sorry, that was a nasty move on my part, throwing that mathematical operator at you.

This exclamation mark is actually saying "**factorial**." What it means is that to calculate the number of route options we have to multiply $10 \times 9 \times 8 \times 7 \times 6 \times 5 \times 4 \times 3 \times 2 \times 1$.

This makes sense because at first we have 10 options to decide from, after visiting the first city there are nine cities left, so we can decide from nine cities for our next destination. We arrive at the second city and we have eight cities to decide from, and so on. The total number of options is expressed in the figure 10!

How much is ten factorial (10!): **3,628,800**

Wow. That is a whopping big number. I wasn't expecting it. ☺

How can you calculate the costs of so many different routes?

No wonder you opted for the easy decision, the local rule of thumb, because if you had to calculate this huge number of options, which is by the way over three million, you would never have boarded the flight.

You would stand at the security check-in, and keep walking back and forth, back and forth, for about two weeks until you calculated the travel costs for all the options.

Even if you had spent the required time on calculations, new information might have surfaced, such as a special discount for a specific travel destination, which would then require that you recalculate all options. **It would offset the calculations.**

The challenge with the travelling salesperson problem is that the number of solutions is increasing exponentially. If you had to visit eight cities, the number of option would have been a mere 40,320. For nine cities, it is 362,880—still below a million. For 10 cities, it is well over a million, as we've witnessed.

By exponentially, we mean that the decision or solution space is increasing rapidly. The increase in the number of possible solutions is rapid.

What happens if your company is increasing in size and has expanded to the entire northwest? You now have to travel to 15 cities.

How big is 15!

It's: **1,307,674,368,000**

OMG that is a huge number!

Again, you have to calculate your least expensive route. No wonder you opt to employ your rule of thumb and decide to travel to the least expensive next destination, hoping it might provide a reasonably cheap approach overall.

However, on your next visit home to Seattle, the vice president of sales has called you in because you spend too much on travelling. He wants to understand **how other salespeople are able to cover the route cheaper**.

Reflect: How is it that others have found better ways to travel a cheaper route?

Can it be luck? Are we even allowed to refer to "luck" as a cause?

You wonder the same. You are also a bit worried. What if your company grows? What if your company is now a nationwide distributor for electronic appliances? How do you decide and plan your travel to minimize costs?

The good old school days

Sometimes I wish I were back at school. Life was simpler then. We didn't have those complex problems to begin with.

The most we had to solve in mathematics were simple quadratic equations. Do you remember those? The quadratic equation that is a second degree polynomial (polynom).

Don't worry, this isn't a curse in ancient Greek. Please keep reading

Polynoms are quite nice and very friendly—especially easy to solve and decide. Businesses would have been much better off solving easy polynom problems.

In school, we've probably encountered polynoms in the second, third, and maybe fourth degree. The math geeks might have seen fifth degree ones.

Agile Decisions

For a second degree polynom such as this

$$ax^2 + bx + c = 0$$

we had a neat solution using a cool formula. You might even remember this formula because you used it for a few exams, and if you remember the formula, you easily aced the exam.

I'm talking about the formula that solves, if possible, the quadratic equation:

$$x = \frac{-b \pm \sqrt{b^2 - 4ac}}{2a}$$

Wow, life was simple back in high school.

This formula is almost like magic. You just plug in the numbers and receive the answer—if there is one.

There are also formulas for third and fourth degree polynoms. However, to solve a polynom, no formula is necessary and even big polynoms can easily be solved by a computer in just a few computations.

Actually, after discussing the traveling salesperson problem, I almost wish that I only had to solve polynoms.

Unfortunately, most of the decision problems we face in the work and business environment, as well as in our personal and social life, are complex problems.

Because the travelling salesperson problem might be viewed as a specific business problem, you might not be convinced yet that complexity challenges our everyday lives.

Therefore, I will share with you yet another complex decision problem. As with the travelling salesperson, it is also unsolvable in reasonable time, if at all.

I'm referring to a complex problem that you're facing every time you go on vacation.

Packing to go the Caribbean

So, there you are living in
New York City—well,
actually, it is Queens, but
still. You're planning an
extended weekend escape
from the city.

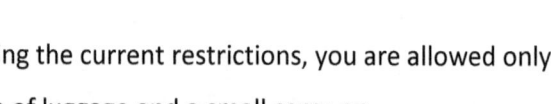

Reflect: What was
your last leisure
travel destination?
How did you pack
for it?

You booked a great vacation
in the Caribbean.

Your flight is early the next
morning. You are packing in
your bedroom.

Considering the current restrictions, you are allowed only
one piece of luggage and a small carry-on.

You have only tonight to decide what to pack. You want to
make sure that you're packing just what you need.

You remember how you forgot a proper mosquito repellant last year on your excursion to Alaska. Who knew the mosquito was their state bird?

Maybe a helmet is in order? Did you know that the common belief that dozens, if not hundreds, of people die of coconuts falling on their heads is no more than an urban fairy tale?

There you are in your bedroom surrounded by your carry-on and bag open, awaiting, and you're looking at your huge closet unsure what you really need to take.

The thing is that each item has a certain utility or, in other words, has a certain benefit. At the same time, each item has a certain volume, or bulk. The transportable storage space is constrained.

You're inspecting the items you'd like to take. You definitely need a pair of running shoes for aerobics, and your second pair of leisure pants just to be on the safe side, your weightlifting shirts that show the biceps, and your moisture repellant cross-training sweater. These are already inside.

You might go snorkeling, so you should pack the lovely snorkeling mask with the fish pattern that was your twenty-first birthday gift that you've always cherished.

The hard copy of the *Silent Influencing* book that you've started reading, with those cool illustrations and creative concepts, is a must.

The checked bag is bursting, and you haven't even packed casual evening dress and an extra pair of pants. You decide that probably the snorkeling fins should stay and that you don't really need your water purification bottle.

 You land in Nassau only to discover that you've left your bathing suit in the top drawer at home, as you were making room for your ukulele...

We all have faced this problem at one point or another and it is a well-known complex problem. Unfortunately, it is not one of those plug-in a formula and get a solution type of problems.

It seems that even the most mundane and seemingly easy problems are really difficult, challenging, and complex. Packing your bag to go on vacation is a problem known as the "knapsack or rucksack problem."

According to Wikipedia:

The knapsack or rucksack problem in combinatorial optimization: Given a set of items, each with a mass and a value, determine the number of each item to include in a collection so that the total weight is less than or equal to a given limit and the total value is as large as possible. It derives its name from the problem faced by someone who is constrained by a fixed-size knapsack and must fill it with the most valuable items.

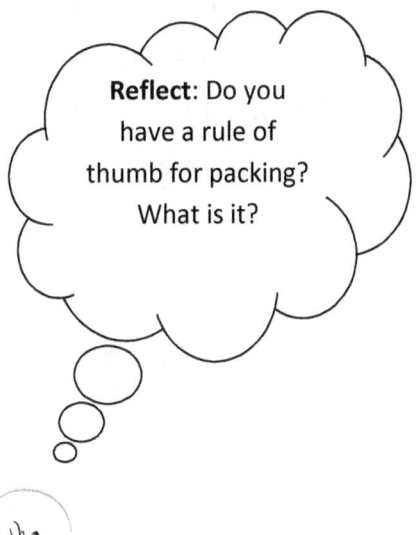

Reflect: Do you have a rule of thumb for packing? What is it?

The knapsack problem has been studied for more than a century, with early works dating as far back as 1897 and further.

It has plagued knights on pilgrimages and middle ages' ships exploring the globe. Due to a shortage in haul space, the ships didn't pack any fruit, which resulted in nasty health problems for the sailors. Only toward the end of the millennium did a British physician recommend packing fruits and vegetables on long sea excursions as a cure against scurvy, which changed the priorities of items to be packed on ships forever.

Summary

There are two types of problems. Ones that are easy to solve known as polynomial problems. We studied about them in school, and sometimes we have formulas to solve them. In general, they are not representative of the challenging business and personal problems we face in our lives.

The second types of decision problems are complex and confusing, and they constitute the majority of business and personal decision problems we face. The number of possible solutions is enormous and grows exponentially with the problem size. Because we can't really calculate the total number of solutions, we opt for rules of thumb for deciding on the best course of action.

Next, we delve deeper into the world of complex problems and illustrate two prevalent approaches to tackle them. We also discover what happens when an airline pilot has food poisoning from sushi in Bangkok, why flying is such a hassle, and whether will we be kidnapped by the NSA.

CHAPTER
TWO

The Sushi was Poisonous

The sushi was poisonous

It is time that we call complex problems by their true name. Yes, complex decisions have a specific name. They are known as NP problems.

NP stands for nondeterministically polynomial.

I thought of adding the definition from Wikipedia here, but I want you to continue reading the book, rather than throwing it in disdain, so I will not include a formal definition.

Suffice to say that NP problems are so difficult to solve that there isn't enough computing power at present, and possibly in the near future, to calculate the optimal solution.

Therefore, we leave the formal mathematical concepts behind in order to investigate and understand more about these complex real-life decision problems and see what we can do about them.

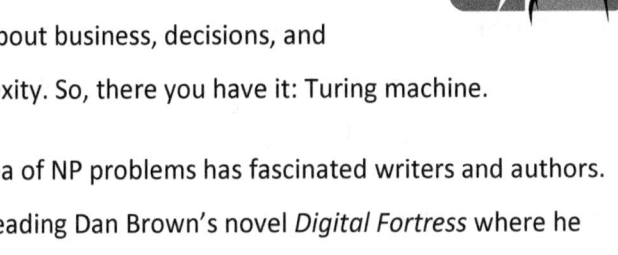

However, I must, at least once, write "Turing machine" for this to be considered a book about business, decisions, and complexity. So, there you have it: Turing machine.

The idea of NP problems has fascinated writers and authors. I was reading Dan Brown's novel *Digital Fortress* where he explained that the NSA already has developed a sophisticated computer that can solve these problems in a reasonable amount of time.

David Baldacci, master of page-turning thrillers, included the idea of NP problems in his book *Simple Genius*. The hero visits a village of geniuses and receives a thorough explanation of what NP problems are and why they are so important.

Basically, the entire digital world around us is constructed on the concept of NP complexity. Encryption is based on the fact that NP problems are not solvable in a reasonable amount of time. In other words, had we been able to solve NP problems quickly, it would prove encryption useless.

Our Wi-Fi network at home and in the office, data communication systems, mobile systems, and secure file transfers are all based on the fact that if you chose a big complex/long password to use for encryption, only the approved receiver could decipher it.

Hence, all the other people, agencies, spies, competitors, and governments cannot decipher your message because no matter how much computer power they have at their disposal, they are not able to crack the encryption in a reasonable amount of time.

Actually, there is a Nobel Prize waiting for you if you can prove you can solve NP complex problems quickly or as quick as you can solve a polynomial problem.

Moreover, if you are able to prove that you can crack complex encryption quickly, not only will get a Nobel Prize, but also $1 million cash.

 You'll also be kidnapped by the NSA and taken to a guarded basement in rural Kentucky to spend the rest of your life among the flying saucer remains from the reported landing in New Mexico and you'll never be heard from again.

Now you understand the reason these complex problems are crucial for our ability to interact, work, communicate, operate, share, and much more in the modern digital network.

We are trudging ahead. Next is a fascinating and enlightening discussion about yet another complex decision problem. We describe two approaches to solve it and journey to Japan, Bangkok, and even ancient Egypt.

The right people for the job

You want to select the right people to do a good job at work, and you want to throw a summer party for your friends and hand out preparation activities to them.

How is that a complex decision problem?

You decide to invite your friends and family to a great summer party. However, you are busy and want assistance in the planning and preparations. Some of the tasks include buying food and drinks, setting up music speakers,

Reflect: Have you prepared a social gathering for many people? How easy was it?

arranging chairs and tables, calling relatives, preparing invitations, mowing the lawn, buying flowers, etc.

Your friends are also very busy and have limited time to help. You're sitting late in the evening, going through the preparations, sending texts, and calling possible helpers. They really want to help you, but their availability is such that they can't perform the time-consuming tasks alone. When you try to pair them up so they work together and reduce the effort of the task, you can't find a time when any pair is available at the same time. You also notice that there are many possible pairings, and it is difficult to analyze all of them, especially given the specific constraints each friend has.

Bob will not go with John, but he will work gladly with Mary. Mary is available on Thursday evenings, which doesn't work for John but does work for Bill. Bill really wants to spend quality time with Bob, and he can work with Mary as long as she can commit to a Wednesday...

You decide to do everything yourself and rethink the value of friendship.

This scheduling problem is comparable to the work-related challenge of assigning people to projects.

At work, you have activities that you perform with others or you are a manager with several people working on different projects. You need to assign the proper people with the right skills set to the different projects. The challenge is that their time is limited, they have various skillsets, the projects have different requirements and due dates, and they have other projects that they're working on.

The known scheduling problem is an NP complex problem. It appears in many business, personal, and social environments.

As we'll examine later, we make allocation decisions for scheduling problems employing a combination of best practices, experience, and usually faulty company-wide procedures.

It is easiest to comprehend the NP scheduling problem in manufacturing production floors because the impacts of

providing sub-optimal solutions are visually seen in the buildup of inventory.

For example, in an auto repair shop, there are several mechanics using a few specific machines to fix cars. You are arriving in the morning to have your car tuned. You want to know when to pick it up.

The manager checks your car and says it requires a tune up and maintenance involving several machines.

Actually, he says that the head cover gasket is faulty and the travel gear shifts are jittery—at this point, you know the invoice will be well over $3,000.

The manager has to commit to a realistic completion time, and the sooner the better. However, there are 15 cars that have to go through different stations and the mechanics differ in skills.

This is a classic job shop environment. In order to provide you with an adequate estimate, the manager must calculate the best allocation of cars to service stations and mechanics, hoping all mechanics show up, no one gets sick, no new cars arrive to the auto shop, and the original inspection of the car provided an accurate assessment of the required repair.

Impossible?

Quite!

Which is why at 1:00 PM you call the manager to receive an update. His response is, "Later, maybe by tomorrow the car will be ready."

The next morning you call and the response is, "Maybe by lunch today because we had some mechanics calling in sick."

You are lucky to receive your car by the end of the day.

Sound familiar?

 Remember that this is the impact of a very complex decision problem in a constantly changing environment.

On production floors, the situation is even more troubling.

Production floors are much bigger than your ordinary auto repair shop. There will be several dozen machines operated and serviced by employees around the clock. The material flows from the raw material warehouse between several machines, ending at the finished goods warehouse.

During its journey, material that has left the raw material warehouse and hasn't completed production is known as "work in progress" or WIP. From the material point of view, **most of its time is spent waiting in queues** to be processed or waiting to meet another material to be assembled.

From a production management point of view, the main challenge is deciding which, how much, when, and at which machine to schedule the processing of the material. They are interested in the optimal usage of machines and the fastest delivery of finished goods, not to mention the minimization of costs.

With several dozens of machines, hundreds of workers, tens of thousands of components, and hundreds of products, this is a complex problem. It also grows exponentially in complexity with the increase of machines, workers, and products. And yet, we can't solve it optimally in a reasonable amount of time.

Does **this** problem sound familiar?

Yes, the characteristics of the problem are quite the same as the travelling salesperson decision problem and the knapsack decision problem. The underlying assumption is that all NP problems are equal, and solving one will actually solve the others.

Thus, if you're able to figure out how to allocate raw material and WIP to machines, you will also be able to easily plan your summer party, know exactly when you'll receive your car, decide which route to take when travelling to several destinations, and know exactly what to pack when going on vacation.

At the same time, you will not be able to use email without others spying on you nor work remotely and send material and payments digitally in a secure manner.  Naturally you will have a million dollars, but also the NSA on your tail.

Given that, at present, we have not cracked the NP decision problem, we'll be using different rules, applications, and individual shortcuts to handle the challenge of complexity.

The two main approaches to tackle complex problems are either top-down computerized algorithms or bottom-up local optimization approaches.

Hence, our next stop is a visit to the United States, where we explore the top-down, computerized approach to solve production scheduling in factories.

Golden arches use computers

The two widely used approaches to solve NP complex problems, be they top-down or bottom-up, were not needed before the industrial revolution or even the middle of the twentieth century.

In the Middle Ages, when you were making a new clock for the king and you promised the king that his new gold and jeweled-encrusted clock would be ready for his fortieth birthday, you had better deliver the clock on time no matter what.

That is, if you wished to retain your head in its same location, i.e., on your shoulders. Of course, if he was a benign king, the repercussions for being late were not as harsh.

 If you were a peasant working the fields, you didn't have a clock, didn't know your birthday, and probably couldn't care less for both.

Starting with the Industrial Revolution and the introduction of machines into production floors, the complexity of planning emerged. Not too much, because most of the machines were the same and the labor was cheap. Planning and scheduling was rather easy.

Things started to become interesting during or after World War II.

At this time, a new generation of machines had appeared, and suddenly there were many, many, many machines able to perform complex operations.

The famous 1908 quote of Henry Ford, "You can have any color as long as it's black," referring to the mass production of the Model T, no longer answered the growing expectations of consumers. These, in turn, were fueled by individualism trends and related marketing campaigns.

Newer products in more flavors, colors, and variations were required to answer people's tastes and preferences. Single machine assembly lines were replaced by complex production floors producing televisions, home appliances, cars, radios, and more. These required sophisticated processing by machines and skilled laborers.

As I've mentioned before, scheduling raw material to production in the described environment is a complex NP problem.

Also, I have previously mentioned that NP problems can't be solved optimally in a reasonable amount of time by computers.

Or can they?

After World War II, computers emerged as the great victors of the war. In the Western world and especially in the United States, a growing appreciation of the calculation powers and the general abilities of computers were evident. Throughout the last 60 years, we have been supporting and believing a narrative in which computers and machines can solve many of the decision miseries of the world. If not today, certainly tomorrow.

It is no wonder that manufacturing and operations managers have **given the NP problem of allocating resources and machines to computers**.

The idea has been that if you give enough information to the powerful computers and let them decide for a few days, you will receive a comprehensive production plan.

Top-down planning has been the general approach to scheduling and managing production.

This is not a bad idea, except when it is.

So, let's imagine how this looks and actually how it still is.

We task our industrial and mechanical engineers with modeling the production floor machine data into the scheduling software.

This data includes:

> BOM: Bill of material, i.e., how much raw material is invested in one finished product;

Reflect: Counterintuitive to our insight, the BOM is very difficult to capture. Can you think why?

> Machine processing capacity; and

> Product processing times.

Once completed, we need to define the required product quantities, which we usually obtain from marketing, hoping their forecasts are accurate.

Friday evening, we hit enter and let the scheduling software run on our server for the weekend. We hope that nothing goes awry; otherwise, we have to visit work during the weekend.

The software application spends the entire weekend analyzing the machines, production capacities, current stock levels, work in progress, and the production forecast.

On Monday morning, the biweekly or monthly production plan is received. It informs us how much and which machine produces what raw material and work in progress. It also conveys restocking recommendations to purchase from suppliers.

The production plan is also known as MRP: Manufacture Resource Planning.

What is the problem with the computer-based decision production plan?

Indeed, I have stated that NP problems **can't** be **optimally** solved in a reasonable amount of time, and this is true. However, I **didn't** claim that the **computer** produced an **optimal** solution!

The computer produced an **okay solution**, i.e., it may be 5% or 7% from the optimal solution. It performed that using specific algorithms.

Agile Decisions

The decision the computer made is based on whatever we've pre-defined. It can be a decision for optimal machine utilization, least usage of labor, or the shortest possible completion time for all products, etc.

A solution that is 5% away from an optimal solution seems adequate. How is it that to this day, production floors are chaotic? There are huge inventories lying stacked in front of machines, and manufacturing departments hardly ever follow through on their promised delivery dates.

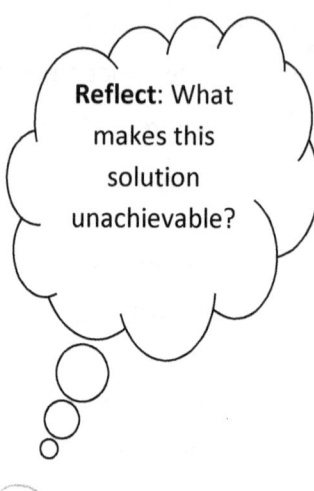

Reflect: What makes this solution unachievable?

Keeping the promised shipping dates is crucial for the continued survival of companies. Consistently committing to delivery dates to our customers is the main reason for investing in complex scheduling software. And, yet, the decision plan fails to deliver products on time.

Of course, one of the reasons has to do with the complexity of the production environment and its **tendency to change**.

Thus, the main drawback with tasking computers to decide on the overall production plan is that the solution they generate is **static**, while production floors are **dynamic** and they constantly **change**.

The plan that took two days to calculate is obsolete after five minutes in production.

Does this sound familiar to those of you working in project environments?

The original decision for allocating resources is irrelevant once you introduce changes; the calculated master plan is rather rigid and cannot endure change.

This is true for production floors, for the project environment, for supply chain planning, for stocking inventory in the supermarket, for a barbershop planning appointments for customers—it is true anywhere where people or machines are allocated across time.

 **Once you have a master plan, you are committed to a rigid static solution that cannot endure change**. This was mentioned already for the travelling salesman problem.

And, yet, the top-down approach for solving or scheduling operations across machines and production floors is popular.

Maybe it has to do with our fascination with computerized and automated processes. We prefer to have a computer make the decision even when we know that the solution will be rendered obsolete once changes kick in.

We agree to pay a lot of money to the IT department to calculate a master plan though we could solve it differently in a more elegant way without the need for expensive software.

In a production environment, it is easy to observe the consequences of the computer created master plan and the workarounds implemented once change occurs.

Production floor mistakes carry visual impacts more than anywhere else because they **translate** to mountains of **inventory**. This is proof that the production plan is no longer valid.

Unforeseen changes, such as machine breakdown, sick leave, changes in forecasts, low quality raw material, and a myriad of various unknowns, render the top-down scheduling plan **useless** for the NP complex problem.

Actually, by default, companies that use these types of planning top-down tools will have inventory buildups because the original plan cannot be maintained—change is inherent in all human activities.

And still, these top down techniques prevail.

I witnessed this scenario numerous times: The CEO and VP of operations go to a sleek tradeshow. They fall in love with the nice graphical interface scheduling application—costing well over $1 million—for planning manufacturing and supply chain. The tool, or rather, the sales and marketing team, promises to solve all late delivery pains.

They purchase and implement the tool, which lasts nearly two years. Unfortunately, even after the application software is completely integrated with the other production floor applications, the top-down plan cannot be sustained, inventory levels are at a record high, and product deliveries are late.

It's important to understand that this is not the result of a faulty scheduling application tool or that people are irrational; rather, it is **because** the underlying decision problem is inherently complex.

Once the original plan changes, what happens? We'll discuss this in the third chapter: building pyramids.

For now, let's hop to the other side of the Pacific to consider another approach to solving complex problems.

Japanese thinking

In Japan, things progressed differently.

Specifically we will be focusing on several manufacturing companies, of which the most prominent is Toyota.

Post-World War II Japanese companies were not allowed to invest in mainframe computers or computers in general. Therefore, they searched elsewhere for solving scheduling challenges in production and material ordering.

In Post-World War II Japan, there had been much interest in contemporary quality management approaches. These were brought forth by two important U.S. quality gurus, who while not highly appreciated in the United States, were embraced warmly by Japanese scholars, engineers, and manufacturing businesses. At the time, the pursuit for higher quality contrasted the still prevailing mass production approach in the west, and specifically in the United States. The lingering perception there was "if you build it, they will buy it," leading to mediocre quality products.

Japan's enhanced production quality enabled clearer visibility of the production floor. Thus, **increased quality enabled better material planning and scheduling**.

Thought leaders in Toyota production were investigating the visual techniques for scheduling and ordering material to and on the production floor. Later, these techniques were labeled "**lean thinking**" and can be summarized as visual manual production communication tools.

 The underlined concept was that we cannot use computers to schedule the entire manufacturing and ordering process, we need to use **local visual signals to communicate** the quantities to be produced.

The idea was to pull products from the end consumer and produce exactly what is needed according to required quantities.

It was in stark contrast to the push mechanism at the basis of the top-down computerized approach driven by marketing forecasts, long production cycles, and mass production.

While the computerized approach is a top-down exercise in solving the entire complex decision problem of allocation of material, resources, machines, and labor to produce the needed products, the Japanese visual pull mechanism is a local decision approach that connects specific product-immediate requirements to the production cycle.

Of note, we are investigating the **specific scheduling** example of an **NP problem** describing **two contrasting approaches** of finding a solution to the complex **decision** challenges.

To a degree, the Japanese decision approach is quite similar to our salesperson rule of thumb, i.e., choosing the next least expensive route. It is what we refer to as an "optimal local" decision.

However, lean thinking was hardly ever labeled as a local decision tool; rather, it was described as a pull mechanism with visual control of production.

How does a lean pull mechanism operate?

Simply described, machines and people produce at their respective station only when they receive a visual signal from someone in a station ahead of them.

It begins with the end consumer. When a consumer requests a certain product, that product is taken from the finished goods inventory, signaling to the assembly station to assemble one product. This in turn signals to stations upstream that they should produce the required quantities.

The signaling proceeds all the way to the raw material inventory warehouse, where components are ordered as they are consumed. This contrasts the computerized approach, where orders are calculated by monitoring monthly levels and yearly sales forecasts.

Lean approaches argue that work in process and waste during production are the biggest enemies of efficiency and effectiveness; therefore, production decisions must be made according to signals from downstream stations.

We're investigating two very different approaches to solve the complex decision problem of scheduling.

One is that we decide according to a top-down calculation and create a plan that centrally issues production orders according to plan.

The second recognizes that things are too complicated to calculate top-down, and anyway, we can't use computers, so it constructs a local level decision mechanism that connects immediate demands to production. It requires a high level of quality performance and employee dedication.

It seems that the local Toyota approach makes a lot of sense.

Do you remember the local rule that we mentioned previously for the traveling salesperson?

Local solutions are easy to grasp. They're visual and alluring because they move away from the complexities of challenging decision problems, giving a set of clear defined guidelines.

As we've already witnessed, these local approaches can be dangerous.

The local decision rule for traveling to the least expensive next destination could result in similar mishaps in a production environment.

Reflect: What mishaps might these local decision rules cause in production?

We discuss more on that in Chapter 3: Flood Management on the Nile.

Taking off: Pilots with Food Poisoning

Taking the lesser of two evils in deciding on NP complexity problems, **lean local optimization pull mechanisms appear to be the better choice**. The global top-down nearly optimal solution doesn't enable **making ad hoc decisions and is inadequate to handle change.**

Therefore, it's logical to use a local rule of thumb.

However, throughout the last seven decades, most industries opted to employ a top-down global approach for business problem solving. Over the last decade, this has gradually been changing.

In our personal life and social life, we tend to favor local rules of thumb. Sometimes this leads to selecting bad options because of our inability to look beyond the horizon and analyze the overall problem.

Are there industries in which a global top-down solution is paramount?

Yes!

Next, we explore the airline industry, which has been implementing top down tools for scheduling aircrafts, routes, crews, and destinations.

Operating an airline is no rose garden, as proven by the number of bankrupt airlines, merged airlines, and completely forgotten airlines that no longer exist.

Yes, once we had TWA and Pan Am.

 What makes operations of airlines so challenging?

For one, you have a limited and expensive resource, i.e., an airplane with a highly trained crew.

You wish to use it optimally.

This translates into flying it from lucrative destinations to lucrative destinations, using it as much as possible, and making sure it is aligned with overall strategy—connecting with various routing options to various destinations.

The number of assumptions and constraints in the airline industry is vast. For example, aircrafts have to be serviced frequently, slots for flights are limited, flight attendants can only work on aircraft they were trained for, and pilots are trained to fly specific aircrafts, etc.

Operating aircrafts in hubs, while making passengers' lives miserable, makes the airline schedule and cost planning easier because logistics are concentrated in one or two main places and the number of possible routing options is reduced.

It is counterproductive to solve the allocation of aircrafts, crews, and destinations without proper scheduling software executed on powerful servers.

In order to create a flight schedule, all the parameters, variables, assumptions, and constraints are fed into the software application.

Remember our traveling salesperson problem from before? For a medium-sized airline that has 3,250 destinations serviced by a few dozen aircrafts and several hundred pilots, the required calculations for deciding on a semi optimal solution is intimidating.

And yet airlines use their software applications to calculate the quarterly and monthly flight schedules.

The challenges arise when specific plans don't follow through.

Imagine that you've traveled to Bangkok for a two-week vacation, and you're supposed to board a flight back home tomorrow.

The pilots who arrived just yesterday to Bangkok with the 747 were having shrimp in coconut with a sushi starter for dinner.

Unfortunately, the sushi wasn't fresh and our pilots had one sake too many—yes, I know that we're talking about Thailand

and Sake is Japanese, they also have quite good sake and Japanese restaurants in Bangkok.

As our pilots had too much sake and didn't see that the sushi was rotten, they woke up at 2:00 AM, and … allow me to skip the details…

However, in one of their seizures at 2:30 AM, they manage to crawl to the phone and make a brief call to the central emergency center back home, notifying them that they won't be able to make the 8:00 AM flight.

You, on the other hand, are totally oblivious to the pilots' condition; rather, you are encountering difficulties in closing your duffle bag and checked in luggage, trying to cram the Thai presents you're planning to carry back home.

The emergency center, in the meantime, executes escalation decision protocol codenamed "bad sushi."

You might falsely assume that they have ample time to reconfigure flight schedules, but they don't. Running the scheduling software takes a few days. In addition, they **can't re-allocate or find another pilot** because all the pilots are booked according to the monthly scheduling plan.

What do they do?

Good question.

If they're a part of a global alliance, they will re-book passengers on other airline's flights, rerouting them to their destination. If the airline is not so fortunate to have a global alliance, they might need to entirely reschedule elements of the plan to make it possible to operate that flight.

Do you remember we said that these plans are very rigid and are not lenient toward change?

The above exemplifies what occurs when you employ a top-down solution for complex decision problems.

A top-down plan is intolerant of change.

I'm sure you've seen that with plans and planning in your business and organization. No matter how big or small your business is or what type of complex problems it encounters, the top-down plan for complex decision problems is highly intolerant of change.

On the other hand, the local decision approach, such as Japanese lean thinking, is more flexible to changes because changes only affect partial elements of the entire problem.

If it doesn't make sense to operate an airline industry in a top-down approach, why do they do it?

In other words, can they actually schedule and plan aircraft allocations in the airline industry with a lean approach?

That's a good question!

Actually up until a few years ago, it had never been attempted. The airline industry believed itself to be too complex; therefore, it must operate around hubs and it has to be tiring, miserable, and overall nasty experience for the traveler.

Can it be any different?

Let's discuss this later when we contemplate fly on-demand approaches. These have started to gain and will be gaining widespread acceptance in the coming years.

CHAPTER
THREE

Building Pyramids

Building Pyramids

Let's observe what occurs once the original top-down plan fails.

What are the mechanisms to solve the NP complexity decision problem once the top-down plan becomes obsolete?

In order to clarify, we will travel back in time to a world without computers, but with diligent engineers undertaking vast projects.

Imagine that you're building pyramids for the Pharaoh. Your name is Khasekhemre. You studied at the engineering school in Alexandria and excelled in your class. You've been building pyramids for the last 20 years and acquired knowledge and experience in doing so.

Khasekhemre: Powerful is the Soul of Re

You're tasked with building a new pyramid for the residing Pharaoh. He's looking for something magnificent, impressive, and huge. Based on your previous experience, this will require approximately 15 years, employing 1,200 slaves and costing 165 years' worth of grain.

You're planning the 15-year undertaking. It consists of preparing the area for the pyramid, dragging huge stone boulders to the site after transporting them on the Nile, carving the stones into blocks of different sizes, building a huge net of bulwarks, and then following the rigorous plan for placing the stones correctly to create the pyramid shape. Apart from slaves, you'll be employing skilled, experienced workers.

Two years into the project, you receive an unexpected visit from the Pharaoh and his trusted advisors. He regrets to inform you that plans have changed and he now requires 400 slaves for a war he's waging with Persia. He also informs you that the pyramid should be ready in seven years to celebrate his 40th birthday.

Going back to your clay hut after a long hot day in the blistering sun, you're rethinking your plans trying to figure out how to complete the pyramids with fewer resources and less time.

Your options are limited: either you elope and head to Greece with the Greek sculptor you've had your eyes on or you produce a smaller pyramid hoping that the Pharaoh will not notice. You prefer not to consider  the third option of not delivering on time as it entails a horrible kind of death.

Following a night of fitful sleep, you wake up resolute to ask for some of your resources back to complete the current plan as much as possible.

You leave the desolate pyramid desert city behind and travel by camel to the Pharaoh's castle.

Reflect: In similar situations, how did you convince your manager, family, or friends to receive more workers, budget, and resources to complete a project?

They who have greater position power

You are convinced that since you've delivered great results in the past and that you've aced the engineering class in Alexandria, the Pharaoh will listen attentively to your arguments and probably favor your request to reinstate the number of slave resources tasked with building his pyramid.

As you enter the great Hall of the Pharaoh, you are disheartened. The war with Persia is led by the Pharaoh's nephew. As you put forth your arguments for receiving more resources to complete the original plan on time, the nephew asks for more slaves from your pyramid work force to use in his campaign.

Actually, you think that you are rather lucky. Two other current undertakings handling Nile control flood zones, which are managed by your fellow engineers from school, have suffered a bigger reduction of resources.

Calculating quickly, you conclude that the Nile flood controls will not be ready for the next season. You make a mental note to

advise the Greek sculptor to move her autumn hut to the hills.

The Pharaoh is resolved to bestow as many resources as his nephew demands.

In the modern organization allocation of people, resources and budgets follow the same pattern. Rather than deciding consistently according to defined decision rules, those with greater position power win.

In the face of a complex problem, deciding between possible alternatives is demanding. The number of variables and options is large and the cognitive effort required big. Those who have greater influence on the decision-maker win the argument and the resources.

The decision is not necessarily optimal from the strategic perspective. The Pharaoh might win the war, but his subjects will drown and the crops will rot.

In the production floor, when the production **top-down plan changes**, as it surely must, an ad hoc decision management team is tasked with allocating the next product to process. This is a daily occurrence in the production environment.

The product manager who has more influence and position power will gain precedent, and the team will instruct manufacturing to work on her product line. From an organization perspective, it might not be the best product to process. However, how can the team know? The decision is too complex. Because they lack pre-defined rules to follow, they decide according to a mediocre rule of thumb that has nothing to do with organizational profits.

Moreover, the rule leads to **shifting priorities**. Once it is clear that position power defines priorities, managers will be calling to instruct the team to change production according to their individual requirements.

At the auto repair shop, we call in to make sure that the manager remembers us. We want to be certain that they prioritize work on our car. Sometimes we consider lingering at the auto repair shop and having another cup of coffee. We figure that our constant presence affects the mechanics and the manager in prioritizing the work on our car so they finish it sooner rather than later.

Similarly, the travelling salesperson might receive a call from a city they plan to visit next month, asking them to visit tomorrow. Since the call is made from an influential office, they concur.

They have no conclusive method of calculating the financial impact of changing the route because there are so many possibilities. This is why conceding to those who have positional power is so alluring.

 On the one hand, we can't provide sufficient evidence that doing otherwise is better, and we are lacking a global rule to guide us. On the other hand, when we can't conclusively show otherwise, following the demands of someone who has positional and influential power makes political sense.

The decision rule that states that those who have more power win the decision is **overpowering and apparent** in many businesses once the original plan fails.

They who yell louder

Yet another decision rule for solving NP complex problems is based on those able to yell louder—figuratively speaking and in practice. It is somewhat similar to the previous mentioned rule; however, it is more damaging because it results in **ever-changing priorities**.

In the Pharaoh example: After conceding to the demands of the nephew, four big projects must squabble over resources.

Your pyramid project is one of them, there are two flood control projects, and the fourth is a new granary for preserving crops.

The four projects are complex, have granular plans, and require intensive manual labor and resources.

The Pharaoh listens impartially to the claims of the four engineers. At this point, he has no preference because he values all four the same.

You, however, are gifted with a very loud and assertive voice and are able to drown the other three engineers' claims by yelling loudest. You assert that without additional slaves the pyramid will never be completed on time. The Pharaoh decides to grant you more resources, leaving the three other engineers subdued and frustrated.

To no avail, the flood project engineer warns that without completion of at least one irrigation network, the capital will be in danger during the coming flood season.

This decision rule, which is based on those who have stronger vocal cords, is probably the **most prominent** decision-making rule to this day.

It's an **aggression** rule that favors those who are more articulate, aggressive, assertive, etc.

It can **wreak havoc on business decision-making** because instead of allocating resources based on value or strategy, it favors allocation based on aggressiveness. It results in ever-shifting priorities given that shouting abilities can be easily developed.

I experienced the "those who yell louder" rule during meetings in a mega refinery project I was consulting with years ago. The total project cost was over $300 million.

Decisions were taken during the daily meetings in which engineers and construction managers participated.

Because the project was extremely complex and the management team **hadn't defined a governing rule to guide decisions** once the plan changed, the managers fought for resources to complete elements of the new production refinery. On any given day, resources and workers were allocated to ad hoc tasks. The complexity of the overall project and the lack of a guiding rule for handling changes resulted in decisions based on the yelling rule.

Surprisingly enough, three months into the project, even the meekest engineers developed amazing vocal abilities.

The yelling rule results in both constantly **changing priorities** and investing effort on **non-critical elements**, thus **postponing the overall completion**.

How does this rule manifest itself in the traveling salesperson problem?

The salesperson might be participating in a video conference call with several of the local managers, inviting their input regarding the next city to visit. In the meeting, the most vocal, aggressive manager warns the salesperson that he must visit his location first.

On the one hand, the salesperson would want to follow the local rule of traveling to the next cheapest destination, but on the other hand, they might agree to travel to the yelling manager's city first.

Since they know that the local rule isn't necessarily optimal, they agree to make adjustments.

What about yelling and the knapsack problem?

The knapsack problem does have several best practice rules of thumb for reaching a near optimal decision. They are based on assigning a relative value to each article to be packed.

However, we will quickly abandon the rule if our mother, spouse, father, son, daughter, or best friend comes barging into the bedroom, handing us an article which we did not consider packing and instructing or demanding forcefully that it be taken along.

Thus, the original decision-making process that we had in place for packing is waived in favor of an external demand. This might be exacerbated if another mother, spouse, father, son, daughter, or best friend comes barging into our bedroom demanding that we pack an item for them—an item we didn't intend to take.

 We might end up travelling with none of our items... It is best to pack as late as possible and refrain from answering calls, messages, or chats before flying long distances to family gatherings.

The "those who yell louder" rule is also evident on production floors. Once the top-down monthly plan changes, the line and production managers meet to discuss allocating and prioritizing product to machines. Naturally the more assertive managers will glean more resources.

The main challenge with decision making for complex problems is that the best solution is not apparent and it is easy to veer off the prescribed course of action.

This is not to say that for easy problems, where the solution is evident, we cannot be veered off by politics and by others who yell louder. However, in low complexity problems (polynomial type) it is easier to observe the impact of irrelevant arguments and politics.

In complex business, social, and personal decision making, where the overall problem is NP, the impact of those behaviors is detrimental and is difficult to trace.

The Pharaoh who decides to remove resources from the flood prevention project to the pyramids project affects his kingdom by considering both his personal preferences and adhering to the yelling/assertive engineer.

Committing to decisions according to the aggressive behavior displayed in meetings **cannot be underestimated**. One only need view the video reenactment of the NASA engineers deciding whether to launch the fateful Challenger space shuttle that blew up minutes after takeoff.

In a virtual team setting, the tendency to follow through on those who speak their minds—ignoring those who are less vocal, less proficient in English, and meeker—can also lead to mediocre business results.

The two rules discussed regarding those who have influence and those who yell louder are not the only personality and power play challenges that affect the decision process.

Next, we present individual **cognitive biases** that affect decision making. As mentioned earlier, these biases aren't specific to NP complexity decision problems; however, complex problems suffer more because it is extremely difficult to observe the consequences of the biases.

Cognitive biases

Apart from being affected by those who have greater position power and those who yell louder, we're also affected by personal cognitive biases when deciding.

There is ample research illustrating that our decision-making process is different from what we'd like to believe.

The fields of research, known as the psychology of judgment and decision-making and behavioral economics, pioneered by psychologists Daniel Kahneman and Amos Taversky, prove our lack of consistency when making decisions.

Reflect: Are you familiar with cognitive biases that affect decision making? What are they?

[Learn more on cognitive bias through Daniel Kahneman's bestseller *Thinking fast and slow*.]

Decision errors, such as anchoring, choosing from the peak and not the average, focusing on the last observed result, etc., affect our ability to consistently decide, especially in complex NP decision problems.

For example, imagine that the Pharaoh recalls that the first pyramid took only seven years to finish. He doesn't remember that the other four pyramids took about 14 years to finish. Thus, he retains seven years in his mind as an anchor reference for pyramid construction time. This, in turn, affects his expectations, the plan he's expecting to review, the number of resources he gives, and naturally, he demands that you commit to the seven years' construction time.

He fails to remember that the first pyramid was smaller or that the Nile was higher and swifter for transferring the building material or any other variable. He only remembers seven years as reflecting pyramid construction time.

This occurs in businesses as well. The marketing department will remember the best delivery date and commit that date to the clients.

On production floors, management tends to remember the best product completion times ever achieved, while manufacturing managers will remember the worst.

Our bias in remembering specific peaks of performance or of past events rather than remembering the averages is detrimental. It leads us to make decisions in complex fields based on hunches rather than on information. Once the original plan changes the guessing game for allocation of resources is based largely on individual cognitive biases.

Apparently, this tendency for cognitive biases is also apparent in other NP complexity problems.

For example, in the knapsack problem, cognitive biases can lead us to remember that packing takes only 10 minutes to complete. This is based on the memory of our previous optimal performance, leading us to invest less time in the effort. It can result in either being late for the flight or, more likely, landing at our destination missing half the things we really need.

So far, we've been discussing what occurs in business,

 social, and personal situations when facing complex NP decision problems. The original plan we developed quickly becomes obsolete, and we adjust based on irrelevant factors that in turn result in poor performance.

What might be the alternative to both solve the complex decision problem and avoid the whimsical decision-making rules that we've presented? How can we protect the top-down plan and yet maintain flexibility as change occurs? What can we do to protect ourselves, at least partially, from cognitive biases? You'll find more in the fourth chapter.

Alternatively, can we select a local optimization approach based on the Toyota way of thinking, protect our team from interferences (political, aggressive yelling, and cognitive biases), and produce according to our local rule, pull mechanism?

Let's elaborate on the possible impacts next.

Flood management on the Nile

The flood control project led by the meek engineer is in a tight spot. On the one hand, the engineer and the project team have to deliver a product that prevents floods from reaching the wheat fields and the capital. On the other hand, they are limited by funding and time.

They have to decide what approach to take. They can produce an unfeasible plan, show it to the Pharaoh, work on it for a few years, and hope it doesn't rain too much during these years—trusting that they will receive more resources in the future.

 This approach is usually evident in modern software and IT projects, and it's probably one of the reasons they usually fail.

The unfeasible plan that modern project teams develop is, well, unfeasible.

It doesn't matter how well balanced the plan is. If it's based on unfeasible assumptions, it will fail eventually.

In ancient Egypt, the engineer and his management team couldn't be so lax with producing unfeasible plans because the personal ramifications would've been disastrous.

What could they do to solve their impediment? They must deliver a product but have fewer resources and less time than the plan requires.

One method is to start working on developing a complete element of the entire plan, delivering it, and moving on to the next. This approach makes sense when the entire flood control project consists of blocking and damming several, usually dry, flood riverbeds.

Naturally, blocking all dry riverbeds is crucial in preventing a big flood from reaching the city, and that was the original plan overall.

However, because the resources are limited, instead of working simultaneously on all riverbeds and providing an incomplete solution on all, it makes more sense to provide a local solution for one or two of the dry riverbeds. This way, the team might be ready in eight months with a complete solution on part of the problem.

Then they can invite the Pharaoh for the completion ceremony and continue later with building the dams on the other dry riverbeds.

This is not a bad solution except when it is. We've already mentioned the drawbacks, and we'll broaden our discussion below.

The local decision-making approach of providing many small complete solutions instead of aiming to solve the overall problem can lead to mediocre results.

In the example of the engineer and the management team building a flood management control system, they must hope that the dry riverbeds they've selected are indeed the ones most essential in preventing floods from reaching the wheat fields and the capital.

In addition, the overall cost of working on many small local solutions might be higher than the approach to working on the complete problem from a top-down approach.

On the other hand, the benefits of working on two or three dry riverbeds might be that, as the work progresses, changes in seasonal flooding patterns will result in floods shifting to other riverbeds. When this occurs, the team can quickly and flexibly move work on those more flood-generating riverbeds.

In project management, we've seen a paradigm change in the last decade, moving from top-down overall problem decisions to local optimized decisions with the advance of **agile project management** approaches.

Much like the flood example, these approaches offer piecemeal, full functional products developed in small steps rather than a full product developed top-down.

We discussed previously the airline schedule decision-planning. We've learned that airlines solve the problem of allocating airplanes and crews across the routes, using sophisticated top-down planning.

We questioned whether this could be performed differently. At present, we are already experiencing a change with the advent of small exclusive fly-on-demand user networks across the United States. Members pay a yearly membership fee and can join point-to-point charter flights. These are still small-scale expensive options, but can the entire airline industry be transformed into a point-to-point on-demand flight service?

In this future setting, you'll be able to book a flight from your city to any other location based on the number of other people wishing to fly to that location. The airline would allocate planes based on demand for a specific travel destination. Thus, you'd be able to **skip flying through a certain hub**. This would be the lean, agile approach to solving the allocation decision problems among the airline industry.

Naturally, this will only make sense with operating smaller aircrafts and having different types of contracts between airlines and airports than we have today.

We've actually already observed this in the traveling salesperson problem.

Reflect: There is one caveat with the local piecemeal decision-making approach. Can you think what it is?

What if the most lucrative route of the airline is New York to London...

However, because the airline doesn't have any available aircrafts—all are deployed for point-to-point destinations and are not available for flying out of either New York or London—we'll be missing the profits of flying this profitable route.

The benefit of the top-down approach is seeing the entire picture.

On the other hand, there are benefits with a local decision approach, as we've mentioned.

Next, we'll examine solutions for combining the top-down approach with the local approach.

Agile Decisions

CHAPTER
FOUR

Fusion, Packing Nothing to Moscow

Fusion, Packing Nothing to Moscow

Top-down decision-making rules have disadvantages, and local decision-making rules have caveats, but what if we decided to take the best of both and combine them?

In order to do so, we would develop a top-down decision approach that views the problem from the overall perspective while allowing for local decisions-making. Our approach would be **flexible to change** while still providing an **overall guiding mechanism**.

Approaches for combining top-down and local decision-making have been discussed in the past. We will investigate some of them, develop guidelines so that you can do this yourself easily, and research the examples we've already presented to see how this would operate in practice.

When everything is important, nothing is

Let's start by examining one of our complex problem examples and observe how the combination of a top-down decision rule and local decision rules can operate.

Once we carry out this exercise, you'll see how intuitive the approach is and how robust the decision process can be. We will then question why individuals, organizations, and businesses often fail to incorporate a top-down decision rule together with local decision rules.

My take on this is that individuals, organizations, and

 businesses prefer to see the drawbacks of a combined approach and **usually opt for an either/or decision-making rule and process**. Philosophically speaking, people prefer to commit to one leading approach because it is difficult to think in terms of a combined rule system.

We'll start with analyzing the traveling salesperson decision problem, which as you remember is NP complex, and the number of route options for a small number of travel destinations is huge. What might a combined decision rule be?

A possible rule can look like this: Before you head out on your route, select three of the 10 destinations that, according to preliminary research, are the **most expensive travel destinations** and calculate the lea**st expensive routes** to these destinations. Once you've done that, use a local decision rule for travel to all other destinations.

The local decision rule remains unchanged: Always travel to the **least expensive next destination** as long as it is aligned with the predefined routes to the three most expensive destinations.

This doesn't necessarily provide the optimal solution, which in any case is not computable. However, it does provide a combination of the top-down approach, which gives a

framework for the decision, together with the flexibility of the local decision approach, which is easy to grasp and easy to implement.

The same is true for the knapsack problem. I'm sure that you've done the following, and it will seem intuitive once we define the hybrid decision-making approach.

For the knapsack problem, we need to perform the following: Select a subset of the total number of articles we're planning to take with us. If, for example, we have 50 articles that we must choose from to pack in our bags, we can decide on the "**must have**" and "**should have**" items.

In Agile project management, we call this decision-making mechanism "**Moscow**": Must have (or Minimum Usable Subset), Should have, Could have, and Won't have.

"Must haves" are product features that must be included before the product can be launched. It is good to have clarity on this before a project begins because this is the minimum scope for the product to be useful.

"Should haves" are product features that are not critical to launch, but are considered important and of a high value to the user.

"Could haves" are product features that are nice to have and could potentially be included without incurring too much effort or cost. These will be the first features be removed from scope if the project's timescales are at risk.

"Won't haves" are product features that have been requested but are explicitly excluded from scope for the planned duration and may be included in a future phase of development.

For the knapsack problem, we need to adjust the MOSCOW decision-making approach by selecting a **top-down rule** that guides us through our packing.

For example, we can decide that we want to make our family happy and, so, all Moscow decisions are based on this guiding principle. Our luggage will be filled with presents to friends and relatives.

We can also decide that we want **never to be cold and wet** on our vacation. Therefore, our **must have** will be a coat, a parka, and an umbrella.

Otherwise, we might wish to be cool and glamorous and pack our best cloths with us as the must haves.

Once you define the top-down decision guideline, the Moscow analysis makes sense.

If you fail to predefine the top-down decision rule, you will not be able to follow through with the Moscow analysis because everything will become a must have.

I'm sure you've witnessed this phenomenon when trying to prioritize tasks at work and at home. You start with a spreadsheet and on the left side are various tasks or projects waiting to be prioritized.

Then, as you meet with your boss, your spouse, or the management team, everything becomes a number one priority.

Naturally, you decide to distinguish between the **really, really important number one priority** and the less **important number one priority** so you **add stars** to designate **differences in number one priority**.

However, in the next meeting everything becomes a **number one top priority with three stars**, making everything the top upmost priority. You repeat the process only to find that you always end with a list where **everything is "number one priority."**

This illustrates what occurs if you fail to pre-define a top-down guiding rule for selecting tasks.

The **Moscow** analysis, while useful, clear, and helpful, will fail if we do not pre-define the top-down approach to guide decision making.

When everything receives top priority, than nothing is important in the sense that **we don't know and can't communicate importance, urgency, and significance**.

The image of the travelers staying up late and hardly waking for the flight because they can't make up their mind on what to pack is all too familiar.

Wearing the same shirt five days in a row

The knapsack problem is challenging when we fail to decide on a top-down planning approach. Airlines have been making life easier, though, by stipulating outside constraints on packing options. By limiting luggage allowance to one carry-on and one checked bag, airlines have actually been helping. This is a counterintuitive byproduct of complex decision problems.

You've seen the unhappy family sitting next to the check-in counter opening all their belongings, sifting through, and re-packing, making sure all suitcases and bags are under the allowed 40 Pounds.

Many times constraints make our lives easier in deciding between options, especially for complex problems. Uncovering the constraints ahead of time limits the number of possible solutions and leads to an easier solution. For example, the traveling salesperson would have **fewer**

options if some travel routes were infeasible or impossible to travel.

If, for example, you communicate ahead of time to your family members that you **never carry gifts**, and you're able to follow through on this rule, you are actually better off packing your luggage. You might also be ostracized and banned from the next Thanksgiving, but this is the price you pay for packing light.

I've met some unruly characters who pack light by committing to wearing the same shirt for five consecutive days when traveling. I don't know how happy their traveling companions are, but this rule clearly makes life easier when packing.

Lee Child's famous fictional character Jack Reacher opts to travel with nothing. Jack's mantra is to have no belongings, no extra clothes, and no extra anything. He's taken the solution to the knapsack problem to an extreme. Because he buys a new outfit every few days, he doesn't have to pack anything and doesn't worry himself with the items he needs to select while traveling.

Constraints make our lives easier. For complex decision

problems, constraints are useful. However, **too many constraints are counterproductive**, rendering the problem too limited.

If you're a basketball coach and instead of having 12 players on your roster you are able to hire only six players, the problem of which players to select for the game is easy. However, you can't make much of a difference.

Organizations that limit themselves can easily find themselves out of business. Organizations that limit their **possible decisions by creating too many stipulations limit their possible solutions and endanger their long-term survival.**

For example, organizations can decide to constrain their portfolio by selecting only endeavors that provide short-term wins.

We've seen this outcome recently in businesses that focus only on the quarterly bottom line. By selecting projects that provide immediate  wins and ignoring endeavors that have long-term prosperous impacts, but take longer to mature, these businesses are destined to fail.

Five Practical Guidelines for Agile Decisions

In order to create the combination between top-down and local problem decisions, here are practical guidelines to pursue that summarize what we've discussed so far and a bit more.

Five practical complex decision problems guidelines:

- Simple local rules
- Strategic top-down rules
- Visual problem view
- Realignments feedback mechanism
- Enforce consistency through publicity

We describe, in detail, each practical guideline below.

Simple Local Rules

This cannot be overstated: **Local rules must be easy to follow**, whether these are rules for a machine operator, a traveling salesperson, a project coordinator, or someone packing bags.

The local decision rules are the ones used most, and they must be easy to follow, understandable, and unequivocal. Consider the warehouse forklift operator who is re-stocking raw material. If she needs to follow a complex decision protocol for placing newly arrived material in the warehouse, it would result in chaos.

Instead, we need to equip her with an easy to follow mechanism for stocking the warehouse.

One such mechanism is **FIFO: First In First Out**. This is also the rule to follow when stocking your fridge with groceries if you don't want dairy products to go sour.

Supermarkets also follow this rule when they're organizing their product shelves. At least, they're supposed to follow this rule; otherwise, they will have outdated products on display.

The modern big supermarkets actually stock fresh items from the back, ascertaining that older items are pushed forward this way.

Sly consumers will then pick up dairy products from the back of the shelf if they want to make sure that it is the

freshest available.

An acronym goes along with local rule simplicity: KISS, which stands for **K**eep **I**t **S**imple and **S**traightforward.

The Japanese Kanban approach previously mentioned implements an easy to follow rule of thumb: when you're operating a manufacturing machine, you should only produce when the physical container carrying a finished product in front of you is empty. In practice, this was actualized by using Kanban cards. Kanban is the Japanese word for "card."

For manufacturing purposes, it much easier for operators to follow this simple rule. However, on ordinary production floors, operators follow a confusing complex weekly/daily production plan. This is the opposite of simple.

Make sure your local decision problem rules are simple

Such as:

- Travel to the next cheapest destination
- Pack the smallest item first
- Work on the easiest element first
- Stock according to FIFO

Strategic top-down rules

The strategic top-down rules we select for our decision problem have to constrain our decision space.

Consider, for example, a hospital's most expensive resource—the operating room. Each such room has to be utilized as much as possible.

Reaching 100% utilization is not feasible; this can be proved using queuing theory. However, 85%-90% utilization is possible and desirable. **The reason we require high utilization is for the pay back on the investment**. The hospital invested money for the equipment in building and equipping the room and wishes to receive a return on the investment.

Hospital operating rooms are generally a resource in shortage; hence, there will always be patients requiring the room. It seems logical that the 85%-90% utilization could be achieved easily. This is not the case. The rooms are utilized at approximately 65% in many cases, which drives the financial officers of hospitals crazy.

It drives them crazy almost to the point that they require ulcer treatment and have to wait in line for the surgical procedure, and yet the operating room is only 65% utilized.

The situation described having lower utilization than expected and desired has to do with selecting an unsuitable top-down problem decision rule. The hospital schedules surgical procedures utilizing the operation rooms by using a monthly plan. As we've illustrated before, the top-down plan fails because of many small changes.

What are these small changes and why do they occur?

In order to perform a surgical procedure in the operating room, the room has to be ready, i.e., cleaned, sterilized, and outfitted with the proper equipment. The surgeon and his staff also have to be ready.

What happens if the surgeon and his staff are ready and waiting, but the room isn't prepped and cleaned?

The impact: We have expensive employees (the anesthesiologist, surgeon, and others on the team), an expensive resource (the surgery room), and a prepped patient—all waiting for the room to be cleaned.

Wait a minute. This doesn't make sense! Didn't the top-down plan specify that maintenance personnel have to be cleaning, equipping, and readying the room?

The plan might have designated and scheduled the cleaning to be performed, but the cleaning staff is currently working at another location and unavailable for cleaning this specific room.

Crazy. We have two expensive resources waiting for important cheaper staff members to finish their tasks.

Why aren't there more cleaning staff members?

Because the top-down rule—that the ulcer-stricken financial officer defined—is based on efficient planning and budgeting of resources. In this plan, it makes no logical sense hoarding on maintenance staff when we can fire them and save...

Thus, selecting the wrong top-down planning decision mechanism leads to an ineffective use of the hospital's expensive resources.

The depicted hospital scenario is quite common in many industries. Eli Goldratt, a physicist by education, claimed that we wrongly select the top-down rules to manage our complex systems and to make decisions. Because the top-down plan will fail, we won't be using the critical resources in our system optimally. He suggested an alternative top-down rule, which he presents in five books. The rule he devised is known as the theory of constraints, and in each book, he applies it to different departments within a company. In each department, the fundamental concept is to analyze the critical resource from the system perspective and utilize it optimally.

The theory of constraints top-down problem decision rule is to **always protect the most limited and expensive resource**, to protect its time, utilization, and allocation.

In the hospital scenario, the top-down rule would translate into constructing the plan around the surgery rooms and expert physicians, making sure that the cleaning staff is always ready, catering to the room.

This approach translates into a seeming surplus of maintenance employees—at times wallowing around the corridors having an extra latte and contributing to the financial officer's ulcer. The alternative though is worse, which is an unacceptable mediocre utilization of operating rooms.

There have been many academic critics of Goldratt's approach. However, most demonstrate that his approach fails in extreme conditions.

For most business purposes, Goldratt offers a straight forward, intuitive, top-down constraining problem decision rule.

An alternative to Goldratt's rule in a production environment and in project portfolio management can be to limit the overall time or products that are processed. In other words, limit the WIP, work in process.

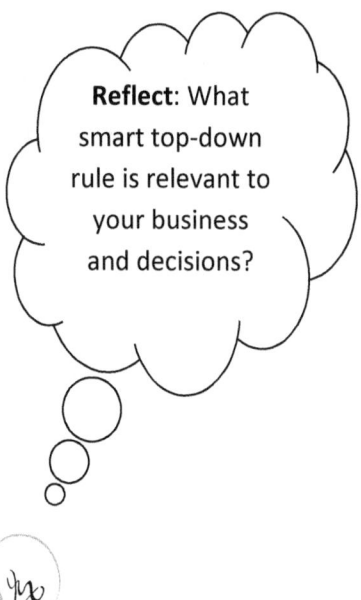

Reflect: What smart top-down rule is relevant to your business and decisions?

How would that rule operate in project portfolio management?

The IT or software departments will only accept new projects when their total WIP is below a certain threshold. The underlying mechanism is of Pull: new projects are pulled into a work status from a backlog queue based on the total number of projects that are concurrently managed at present.

The constant WIP is an easier rule to manage, but it is difficult to determine the threshold and commit to it without surrendering to requests from top management. More on that later.

Select a resource constraining top-down rule, instead of planning the entire system.

Visual Problem Presentation

"If we don't see it, it's not there."
We assess the world around us through our eyes.

"A picture is worth 1,000 words."
The fundamental principle is that we would rather see a visual representation rather than a text list.

"Seeing is believing."
Philosophically speaking, visual presentations sometimes can lead to mistaken results; however, more often, they DO enable us to clearly see the overall problem.

 Because we use our eyes as the tool to capture the world, it is crucial that we inspect and evaluate decision problems in the same way.

Visual representations of decision problems enable us to better grasp the problem and assumptions, constraints, options, and solutions.

Reflect on the traveling salesperson problem. It's much
easier to explore a tangible
map of 10, 15, or even 30
destinations and analyze
possible routes than to assess
the problem using a table or
spreadsheet.

Reflect: How can
you present the
top-down rule on
the visual
representation of
the travel route?

The same is true for
production environments. It is
easier to solve the complexity
of production machine
allocations using visual
signals, such as Kanban
boards and cards, rather than

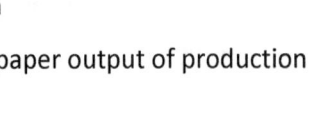

using a computer-generated paper output of production
orders.

One of the challenges we've witnessed in managing multi
projects in virtual global companies is that we lacked the

visual representation of the activities and projects allocated to resources and people.

 **Use any relevant tool to display the decision problem visually.**

For example, during a project at a petrochemical plant, we created a war room for the intense four-month construction stage. While we had a detailed Gantt chart with over 2,000 entries for specific activities of construction, it made more sense to create huge visual boards depicting daily tasks with resource and people allocated to display overall allocations, possible collision points, impacts, deadlines, opportunities, and threats.

The strategic top-down rule has to be shown in the visual representation of the problem. Hence, it is not enough to portray the problem using a visual approach, it is also important to superimpose the strategic top-down rule on the visual presentation of the problem.

How can we illustrate the strategic top-down rule on the visual image?

- Using computers, this would be done adding an extra layer on top of the decision problem.

- In our physical war room example, this was actualized by different colors, stickers, and other visual tools to depict the strategic decision rule visually.

- On production floors, this would be achieved by adding physical signals and drawing signs and images around and in front strategic human operators, machines and other facilities within the production floor.

Create a visual representation of the complex decision problem. Use visual methods to illustrate the top-down strategic decision rule.

Realignments feedback mechanism

Integrating the top-down strategic decision rule with the local simple decision rule is the first step in analyzing our complex decision problem.

Once this has been achieved, we use local rules to manage the daily and weekly decisions. **The challenge is that we might be steered away from our strategic objectives**.

Remember the Egyptian flood control project? The meek Egyptian flood control engineer can start the project by damming the proper riverbed to prevent floods. However, as work progresses, the team can find itself moving to work on an adjacent dry riverbed that seems to affect flood levels.

Unless the engineer and the team invest time to realign their plan with the overall flood management plan, they might be working in the wrong place. After all, while the adjacent riverbed affects overall flooding, it isn't the most impacting riverbed, and thus, they will be wasting effort and time.

This decision-making drift is common in long endeavors and must be acknowledged and tamed.

It is very similar to the local problem decision rule that we've analyzed previously for local allocations of aircrafts. Lacking an overall alignment mechanism, which ascertains that we always operate an aircraft flying the lucrative route from New York to London, we can easily find ourselves missing the required resources for our most profitable route.

 The challenge is that since local decision rules are very easy and simple to operate, they can lure us into ignoring the bigger picture and overall strategic objectives.

Take, for example, the Japanese pull mechanism for managing production floors. The local rule instructs machine operators when to start producing the next item.

The system is based on pull from the end consumer, to the finished goods warehouse, to assembly, production, and all the way up to the raw material inventory.

The decision rule is simple: An operator needs to produce only when there is a card or an empty container in front of him, signaling that the next station requires the produced item.

This doesn't guarantee that the production is actually aligned with the strategic top-down objective. Naturally, when production started manufacturing more than a year ago, the pull and signaling rules probably answered strategic need.

However, at least on a monthly basis, we must maintain a realignment procedure, ascertaining that the local Kanban pull mechanism is aligned with the overall most profitable product.

Thus, when using a local decision rule, such as Kanban— which by the way is extremely visual—we must realign our daily and weekly plans with the overall plan every so often.

This is true whether we are using Kanban pull signals for production or for Agile project management.

In production, it is better to realign our decision mechanism before products are processed by the strategic resource.

These can be machines or people. Much like the operating room in our previous example, strategic resources determine the profit we are able to extract from the process.

Hence, it's preferable to realign our problem decision rule with the top-down rule before items go into strategic resource processing. **It would prevent spending expensive strategic resource time on non-critical items**.

To achieve it, we would examine the waiting queue that usually builds before an item goes into strategic processing. We can inquire whether the specific waiting item is indeed aligned, with the overall strategic objectives that we've defined.

The misalignment occurs often in Agile project management when producing software, specifically user interface design.

The intensive user-developer daily and weekly exchange can quickly lead the overall user interface to functionality, which is not relevant for the overall user community. Without a monthly or quarterly realignment process, the team will be wasting efforts on producing a product that is not aligned with overall strategic objectives.

The realignment feedback mechanism has to be built into the decision process. The traveling salesperson, moving between 30 destinations, needs to reanalyze the planned route every six or seven destinations.

If, for example, the most important destinations were initially identified as strategic and most profitable, this needs to be reevaluated when 25%, 50%, 75%, of the overall destinations have been reached. The local rule of  traveling to the least expensive destination next should also be reevaluated because alternative simple rules might be preferable at this point.

When packing our luggage for vacation, we would probably like to examine what we packed based on our strategic rule and local decisions and analyze whether indeed we have followed our local and global rules. We'll likely find that at some point during our packing process, we veered away from the prescribed decision rule and started packing things based on ad hoc irrelevant proclivity.

Make sure you build realignment decision points into your complex decision process. The rule is not too many and not too few, depending on the problem size and duration.

Enforce consistency through publicity

The most emphasized complaint I've been hearing from junior and senior stakeholders, following through prescribed decision-making protocols, is their disability to follow through on prescribed decision-making protocols. ☺

We've mentioned similar effects when discussing prioritization of items on a list. Usually, all items become upmost priority. This also occurs when managing business risks; all identified risks quickly turn into top priority.

How can we ensure that we follow the prescribed decision process?

How can the traveling salesperson not succumb to late-night urgent phone calls from angry vendors residing in cities not strategic and remain faithful to his prescribed route?

While packing our bags, how can we decline a request from family members, to carry two bottles of wine as a gift to our distant relative?

How can manufacturing engineers refrain from prioritizing products just because they received an angry reprimand from a senior executive?

These questions are probably the hardest to answer. All too often, I've seen robust scheduling decisions that had been carried out from a top-down perspective considering strategic resources eradicated by powerful stakeholders.

While this book provides a prescription for deciding how to face complex decision problems, it would not be complete without detailing a powerful and practical tool to use, **ensuring that in practice, we can carry out our top-down and local decisions**.

This tool has been described in Cialdini's *Influence: The Psychology of Persuasion* as commitment and consistency, which claims it is easier to be consistent with things to which we're committed.

According to Cialdini, we like to be consistent and honor our commitments. The famous experiment was the one in which a research accomplice goes to the beach, lays down on a blanket, and puts out some personal items, including a radio. The accomplice then goes away.

A few minutes later, another accomplice comes up and "steals" the radio.

The experiment varies between two conditions.

In one condition, the original accomplice gets up and leaves without saying anything.

In the other condition, the original person specifically asks someone to watch his or her stuff.

In the first case, four in 20 times, the second accomplice could "steal" the radio without challenge.

In the second condition, the "thief" was stopped and challenged *19 out of 20 times*.

So the challenge rate went from 20% to 95%. **In other words, people had an overwhelming desire to be consistent with their prior commitment.**

These techniques can be remarkably subtle.

For example, when a telemarketer calls, it makes a huge difference whether he says, "How are you feeling tonight?" and gets an answer or says, "I hope you're feeling well tonight."

The difference is that in the first case, the target has **committed publicly** to having a good evening (because the response is typically, "Fine, thanks"). Having publicly committed to doing "fine," it is very hard for the target to avoid giving money to the earthquake victims who are obviously not fine and in need of help.

In the case where the caller simply says, "I hope you're feeling well this evening," **no such commitment was extracted and the response rate was less than half** (15% versus 33%) compared with what it was when the caller did ask a question.

Another example is with toy companies that advertise items in the run-up to Christmas, which they have no intention of stocking in sufficient numbers to meet demand. The unwitting parent **commits to the present for the pleading child**.

Because the gift isn't available, parents purchase anything of equal value for Christmas.

However, two months later, the original item is stocked on the shelves. The parent buys it because he feels a commitment to his kid.

Toy manufacturers are aware of this commitment and use the technique to prop up sales in January and February.

Companies use essay contests to make you feel good about them. Something as simple as copying out a message in your own handwriting convinces you to follow through on all the nice things you've said about that company.

Public utilities have gotten people to save energy simply by convincing them to commit to saving energy.

It turns out that internalizing the commitment is key.
When utilities hold a contest and say that those who save a certain amount of energy will be recognized, people cut down on energy usage. But when they then call to renege and say the contest is cancelled, it turns out that energy usage *falls even further*.

It seems that this is because people are actually less motivated when they feel they're doing it for external reasons **and more motivated when they feel they are doing it for themselves**.

Being the kind of person who likes to be energy efficient is more powerful than being the kind of person who will reduce electricity usage in order to save $5.

By now, you're convinced that commitment and consistency are powerful influencing tools.
How can we use them to follow through in decision protocol?
We must **commit** to our **decision-making protocol**—the one we described previously that includes a top-down strategic and a local simple decision-making rule—in **public**.

The commitment should be in writing and communicated throughout the business. We have to make sure that others **know** that we've publicly and formally committed to our process of decision making.

By doing this, we can reply to pesky stakeholders that we'd love to help them, but the formal committed process requires us to perform differently.

Of course, this is not a foolproof approach, but it provides substantial robustness for the decision-making process.

 **It is recommended to embed the decision problem rule in the overall process to ensure commitment.**

As an example, reflect on the Kanban approach:

It is built in such a way that it would **withstand senior management stakeholders asking for prioritization of specific products**, as the Kanban container or card system provides a formal commitment of production.

This is also the case in Agile project management when using visual Kanban boards. These physically present the allocation of people's activities across time. It would be difficult for stakeholders to change this visual formal commitment of allocated work.

The same is true for the traveling salesperson who communicates publicly his plan and submits it along with a predefined process for making changes.

The first four practical rules are useless without a social, business mechanism that commits the parties to a consistent problem-solving protocol.

Make sure you commit to your decision-making process publicly and formally and communicate it.

Michael Nir

CHAPTER
FIVE

Decisions: the Road not Taken

Decisions, the road not taken

TWO roads diverged in a yellow wood,
And sorry I could not travel both
And be **one traveler, long I stood**
And looked down one as far as I could
To where it bent in the undergrowth;   5

Then took the other, as just as fair,
And having perhaps the better claim,
Because it was grassy and wanted wear;
Though as for that the passing there
Had worn them really about the same,   10

And both that morning equally lay
In leaves no step had trodden black.
Oh, I kept the first for another day!
Yet knowing how way leads on to way,
I doubted if I should ever come back.   15

I shall be telling this with a sigh
Somewhere ages and ages hence:
Two roads diverged in a wood, and I—
I took the one less traveled by,
And that has made all the difference.

Robert Frost, "The Road Not Taken"

Agile Decisions, satisfying rather than optimizing

In *The Paradox of Choice: Why More is Less*, Barry Schwartz discusses how people can be broken down into **maximizers** and **satisfiers:**

- **Maximizers** try to find the best possible option of all possible choices.
- **Satisfiers** know what their needs are and choose the first (or any one of the first) choices that fulfills those needs.

On the surface, it may seem that the maximizer has a better strategy. By carefully analyzing the situation and weighing the options, he tries to make the absolute best choice.

 Ergo, the maximizer always makes the best choice, right?

Maybe not.

As it turns out, over time, the **satisfier is the one who tends to do much better**.

Why is that? Well, for a few reasons.

The maximizer is more prone to paralysis if he can't decide between two options that are both strong in different ways. The maximizer spends (i.e., wastes) a lot more time and energy finding all possible alternatives and then trying to decide which option to use.

The maximizer spends more time afterward questioning the decision he made.

The satisfier, while not always making the absolute "best" choice, makes **quicker decisions**, doesn't become paralyzed, and tends to be happier with his choice afterward.

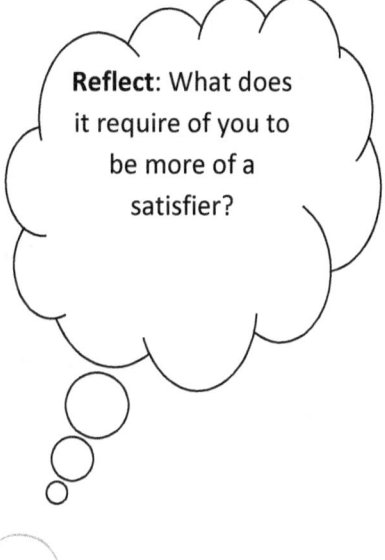

Reflect: What does it require of you to be more of a satisfier?

Focus and Limiting Options

In this book, I presented the **scientific basis supporting the satisfier approach to decision making**.

The challenge the optimizers face is their need to completely weigh all the possibilities.

In light of NP complex problems, we understand that this is a fallacy.

As I am drawing to the summary of the book, I want to share with you the essence of effective decision making, which is one word:

Focus

Focus is the one skill that effective, satisfying decision makers possess.

The complexity surrounding us is staggering.

We have issues with decision making not because we're irrational, but because the problems we're facing are truly complex.

We need to **focus on the complexity and embrace** it.

Many times, we substitute the complex problem with an easier one that does not reflect the complexity of the original problem.

We need to focus on the **elements of the complex decision.**

Other times, we let a computer solve the problem, which as detailed in the book, doesn't provide robust results that can endure change. We pay a hefty price because we believe the computer provides the best solution.

We must **focus on our problem-solving skills and accept the impacts of change**.

There are those who do not yield to the complexity and ask for **more and more, and yet more, data** to support a decision, not understanding that it is a futile quest. Complex NP problems are huge and beyond our cognitive ability to grasp.

We must **focus on the time limit we have for deciding**.

Decision making places us between a rock and a hard place. Psychologically speaking, the decision process taxes our mental and physical stamina because, when deciding, we are losing degrees of freedom. We are losing possible options and choices. This is difficult for us to accept. Robert Frost's poem, which I've included above, subtly relates to the diminishing degrees of freedom.

And sorry I could not travel both.

Accepting and letting go is crucial.

This is the hardest thing to do, but probably the most important. Once you make a choice, let it all go. Stop worrying if you made the right choice. Stop obsessing over what you might be missing. And don't trouble yourself wondering what life would be like if you took the other fork in the road. You can't go back and change the past, so just let it go and enjoy the present.

Ultimately, this is one of the key things that separates people who make satisfying decisions from those who don't. If you know you will be able to be happy with your choice and let it go afterward, you relieve a lot of the pressure and fear of making the decision. If you are worried that you will be regretting your decision for days, weeks, months, or even years, you will feel tremendous stress and go into over-analysis and paralysis.

Focus: Need, top-down rule

Remember to focus on a top-down rule/top-down mechanism that guides your decision-making process.

It is crucial no matter what the decision is.

Often, we are confused with the sea of options before us, and we fail to distinguish between needs and wants.

Sure, you want it all. But what are true needs and what are just wants?

Needs are things that you absolutely **must** have.

These are the basis of the top-down strategic rule, such as **high utilization of surgery facilities**.

Wants are things that are nice to have, such as **reduce the maintenance staff cost**.

Problems arise when people don't think through what their wants and needs are. As a result, they become sucked in by cool "wants" and end up going with a choice that lacks one or two of their "needs."

Another challenge is mistaking "wants" for "needs." It's not uncommon for a person to come up with a "must have" list that has 20 or more items on it. That's crazy. Most of those items are probably "wants," but the decider is convinced that they are "needs."

As a result, they become paralyzed setting out on the near-impossible task of satisfying a massive "needs" list.

For most of the choices you make, big or small, you should only have a few needs. Three is usually plenty.

For example, if you are considering buying a house, you may have a huge list of "wants," but how many of those are truly "needs"? You could probably narrow it down to a handful: minimal repair issues, design, and a good school system/neighborhood.

The need guides the decision-making process and it should be the defined guiding top-down rule.

Michael Nir

SUMMARY

What is unique about this guide

In the previous chapter, I mentioned the five practical guidelines of Agile decisions.

Effective decision making is about **being true to a predefined process.**

Due to the taxing psychological effort that decision making extracts, the process protects us from falling into the many traps that complex decisions present.

The practical guidelines presented, having a guiding top-down rule—a need—supported by local simple rules for "small decisions" using a visual representation, periodically realigned, and enforcing consistency, are an extremely powerful method.

Yet I haven't seen it much in literature.

My aim in this book has been to combine several areas of knowledge seldom viewed as integrated.

- Decision-making books often focus on human shortcomings and cognitive weaknesses. Others present a tiring collection of procedures, templates, and processes in a rather admonishing approach.

- Complexity and NP problems are discussed in computer technology, scientific tomes. Truly, we can profit from appreciating the inbuilt complexity of real world problems, and allocating people to tasks is NP.

- Business books, specifically, present scheduling in manufacturing and projects as discreet topics not related to the overarching decision-making complexity.

- Lean manufacturing and Agile project management have hardly been described as local decision-making approaches, which misses the explanation of the intrinsic benefits and drawbacks.

Finally, I want you to think of complexity the next time you

 pack your bags for either on a business trip or on vacation. Think about landing at your destination only to find you forgot your battery charger, however, we are not predictably irrational.

We are not bad decision makers as many books tend to blame us; rather, the problems we face are inherently complex.

I offered my take, one I have been using successfully in consulting and coaching. I hope you make the best of it.

What's next and final words

This is my ninth book.

To a degree, it is different from most of my other titles. It follows a keynote I presented on the topic of portfolio decision making in business.

I did mention some decision-making concepts in both of my Agile project management books within the context of Agile project local decision approach and the drawbacks are inherent.

My other titles also include illustrations and thinking alerts, the cornerstones of modern short books.

First, *Influence and Lead* is always available FREE as a Kindle eBook.

Feel free to download it from the U.S. Amazon site.

Influence and Lead ! Fundamentals for Personal and Professional Growth

In addition, Building Highly Effective Teams is an all-time bestseller that receives great feedback from readers.

One of my personal favorites is Silent Influencing, a definite must read, which includes comic illustrations such as this:

If you don't eat your meat, you
can't have any pudding. ...

All titles are offered as print editions as well, and usually offered at a substantial reduced price by Amazon.com.

I really hope I've succeeded in making this decision-making book an enjoyable and interesting adventure for you. This book illustrates the complexities of what decision making is about and is a fun and easy to use practical guide for you to reference as you're developing a decision-making approach.

Hopefully you've enjoyed it and I'll meet you in my next book.

Thanks again for your time. I appreciate your interest and focus.

Michael, 2014

www.ingramcontent.com/pod-product-compliance
Lightning Source LLC
Chambersburg PA
CBHW051503170526
45166CB00001B/369

* 9 7 8 1 5 0 0 5 3 1 0 7 2 *